Carter's Grove

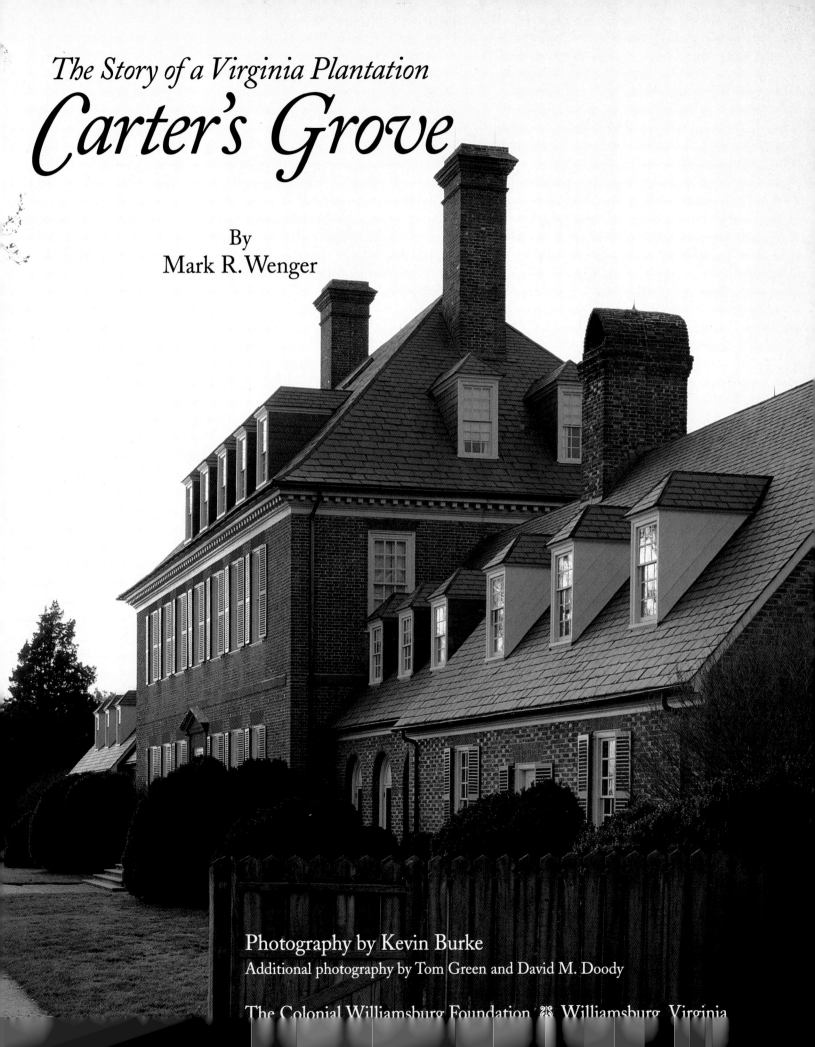

The Story of a Virginia Plantation
Carter's Grove

By
Mark R. Wenger

Photography by Kevin Burke

Additional photography by Tom Green and David M. Doody

The Colonial Williamsburg Foundation ❦ Williamsburg, Virginia

Library of Congress Cataloging-in-Publication Data

Wenger, Mark R.
    Carter's Grove : the story of a Virginia plantation
/ by Mark R. Wenger; photography by Kevin Burke.
        p.   cm.
    ISBN 0-87935-129-2
    1. Carter's Grove (Va.)   I. Burke, Kevin. II. Title.
F234.C35W46 1994
975.5'4251—dc20   94-26980
                                              CIP

Designed by Helen Mageras

Printed in Hong Kong

The Land and the People

*I*n 1619, about two hundred English settlers sponsored by the Martin's Hundred Society established a new community they called Wolstenholme Towne on the bank of the James River. Named for the company's principal stockholder, Sir John Wolstenholme, the fledgling outpost was the administrative center of Martin's Hundred plantation, a vast tract comprising nearly twenty thousand acres in what is now James City County.

These English colonists encountered a thriving Native American culture. For thousands of years the Powhatans and other tribes had extracted their living from this land and from the watercourses that surround it.

Sometime around 6500 B.C. small groups of Native Americans who traveled throughout the year established several campsites at Carter's Grove. Artifacts recovered from these camps include stone spearheads, knives, specialized hide working tools, and flakes of stone created during tool manufacturing.

With the introduction of agriculture about two thousand years ago, Native Americans began to live in more permanent villages. Each spring and fall the villagers broke up into smaller hunting and gathering groups that exploited seasonally available resources within the region. The presence of large collections of oyster shells and broken pottery dating from A.D. 300 to 1000 which have been found clustered around the ravines that cut through the Carter's Grove landscape suggests that Carter's Grove was a frequent destination of these small groups of hunter-gatherers.

The site of Wolstenholme Towne's fortified compound is marked by a simplified reconstruction. Beyond it lies the James.

In 1585, John White made detailed drawings of Virginia's native population. Working from one of White's sketches, Theodore de Bry produced this engraved view of Secota, a village in what is now North Carolina.

De Bry identified certain elements of the drawing as follows: [A] The tombs of their kings and princes; [B] Where they assemble to make their solemn prayers; [C] They have broad plots where they celebrate their chief solemn feasts and [D] a place where after they have ended their feast they make merry together; [E] They have gardens wherein groweth tobacco which the inhabitants call *uppowoc;* [F] In their corn fields they set a cottage like to a round chair wherein they place one to watch; [G] The leaves of their corn are large, like unto the leaves of great reeds; They sow their corn with a certain distance noted by [H] otherwise one stalk would choke the growth of another; [I] They have a garden where they sow pumpkins; [K] A place where they make a fire at their solemn feasts, and hard without to the town a river, [L], from whence they fetch their water.

Although the site has not yet been found, sometime after A.D. 1000 at least one large village appears to have been located at Carter's Grove. Crops grown near this village would have included corn, squash, and beans. Within the village itself there would have been semi-permanent homes, storehouses, work areas, and space devoted to religious activities.

But this way of life could not be reconciled with the goals of the European settlers. Alarmed by the growing frequency and extent of English encroachments, on March 22, 1622, the Powhatans rose up in an effort to destroy the interlopers at Martin's Hundred and other plantations along the James River. When the onslaught was over, half of Wolstenholme Towne's inhabitants were dead or missing and virtually all of its buildings lay in ruins.

The surprise attack had been a tactical success, but it could not reverse the rising tide of English occupation. Throughout the remainder of the century, white settlers managed a tenuous existence here, dividing the land among several small farmsteads, all devoted to the culture of Virginia's staple crop, tobacco.

Though Africans arrived in Virginia as early as 1619, they were few in number at that time and probably shared the status of white indentured servants, who in return for paid passage to Virginia bound themselves to a master for seven years. Not until the 1640s was the institution of slavery fully established in Virginia law. Indentured whites were initially the principal source of labor for Virginia's tobacco-growing enterprise. By 1700, an acute shortage of these indentured servants led tobacco planters to look elsewhere for the labor they required. A booming trade in African slaves was the sad result. Within a generation, black laborers replaced white servants in the tobacco fields of Virginia, and slave quarters like the one now reconstructed at Carter's Grove soon dotted the landscape.

The reconstructed quarter stands on the site of a late-colonial slave dwelling.

During the first decades of the eighteenth century, African slavery stabilized the labor supply and rising tobacco prices ushered in an age of unprecedented prosperity. In the process, a handful of interrelated families rose to social and political dominance. Among the ascendant tobacco planters, none was richer or more powerful than Robert "King" Carter, owner of 300,000 Virginia acres and more than a thousand slaves when he died in 1732. Carter's stupendous holdings included the Martin's Hundred tract. In the will drawn shortly before his death, Carter left this property to his daughter, Elizabeth Carter Burwell, and her eldest son, Carter Burwell, stipulating that "in all times to come it be called . . . Carter's Grove."

Two different worlds . . .

Remains of Wolstenholme Towne's fortified enclosure, excavated in 1977.

Carter Burwell made his residence here by 1738 and soon began construction of the mansion complex, a project that took about fifteen years to complete. Burwell's magnificent house was the product of a flourishing consumer society made possible by Virginia's expanding wealth and growing access to credit. Looming over the outbuildings that clustered around it, Burwell's great house aptly portrayed his exalted place in Virginia's colonial society.

As a revelation of the builder's personal qualities and ambitions, Carter's Grove was equally revealing. A remarkable structure for its time and place, it surpassed all but a handful of its contemporaries in size, elaboration, and classical rigor. At the very time when the mansion was under construction, Burwell mounted a campaign to obtain a seat on the governor's Council, the upper house of the Virginia assembly.

There is, perhaps, no clearer evidence of what this structure represented in 1755 than the contrasting worlds of Wolstenholme Towne, with its meandering lines of postholes huddled on the river's edge, and Carter Burwell's great mansion, standing assertively atop its manmade terraces, dominating everything around it. Nowhere are the builder's self-concept and attitude toward the land more dramatically displayed.

Burwell moved into the new house in 1755, but his enjoyment was short-lived. He died in the spring of 1756, leaving a widow and nine children, including his eldest son and heir, six-year-old Nathaniel.

Not until Nathaniel came of age in 1771 did Carter's Grove again become the Burwell family's primary residence. Returning to the great brick house on the James, Nathaniel made his home there for the next twenty years.

Nathaniel Burwell, son of the builder, as he appeared late in life. *Courtesy, Clarke County Historical Association.*

Overleaf. Enlarged and modified in this century, Carter Burwell's mansion stands atop manmade terraces hewn out of the land nearly 250 years ago.

After the end of the Revolution, eastern Virginia entered a period of decline that saw many leading families move westward in an attempt to sustain their fortunes. In 1792, Nathaniel Burwell shifted his residence to the fertile lands of the Shenandoah Valley, where he built a new house, Carter Hall. His heirs sold Carter's Grove in 1838, and the estate then passed through the hands of numerous owners in a long succession of auctions and foreclosures.

Very little happened to the house in the way of major changes until Edwin G. Booth purchased it in 1879. Booth was a "New South" man, an enthusiastic supporter of railroads and new industry as the keys to rebuilding Virginia after the Civil War. Booth tirelessly promoted reforms of every sort, but most of all he sought to encourage a genuine restoration of the Union. In 1881, Americans celebrated the one hundredth anniversary of George Washington's victory over the British at Yorktown. Booth saw the Yorktown observances as a fitting occasion to celebrate the theme of reconciliation. In commemoration of the event, he painted the mansion's interior red, white, and blue, and planted a grove of locust trees on the landside approach, a popular form of commemoration during the Centennial of American Independence. Elaborate porches were added to the front and rear of the house for the sake of convenience.

The river-front entry as it appeared following Edwin G. Booth's red, white, and blue paint job.

Top. The Booths added new porches front and rear. By this time the tulip poplars on the river front were already quite large. *Courtesy, Valentine Museum.*

Center left. The new land-front porch. *Courtesy, Papers of Hugh S. Cumming, Manuscripts Division, Special Collections Department, University of Virginia Library.*

Center right. Edwin G. Booth—lawyer, philanthropist, New South prophet.

Bottom. Carter Burwell's dining room was now almost 150 years old. *Courtesy, Joseph Tilden.*

Leaving the Booths' river-front veranda intact, the Bislands added exterior shutters and, at the extreme right, a gallery connecting the kitchen to the main house. *Courtesy, Valentine Museum.*

Modern amenities were introduced in 1907 when Carter's Grove became the property of T. Percival Bisland, a New York businessman with ties to the silver mining industry. Assisted by New York architect William W. Tyree, Bisland rehabilitated the mansion, adding such conveniences as window screens, indoor toilets, central heat, and a modern kitchen in one of the flanking dependencies that was joined to the house by an enclosed gallery. Like Carter Burwell, the new owners found little opportunity to enjoy the house on which they had lavished so much energy and care. By the fall of 1910, both Mr. and Mrs. Bisland were dead. Once again, Carter's Grove languished in the hands of absentee owners.

"Horseplay" with canine friends.

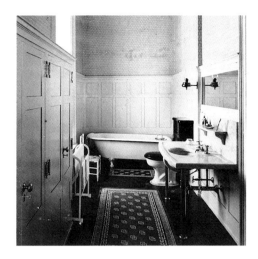

Above, left to right. A pantry and scullery occupied the new connector between the house and kitchen.

A refurbished upstairs room.

Modern amenities—indoor plumbing on the upper floor.

Center. On the land side, the Booth-era porch was removed and the old steps returned to their original location. *Courtesy, Mrs. S. H. Ross.*

Left. This unidentified family lived on the property and shared in the life of Carter's Grove. The man in white worked in the house as a Bisland employee.

In 1928, Mr. and Mrs. Archibald McCrea purchased the estate. McCrea, a Pittsburgh industrialist, had married Petersburg widow and society belle Mary Corling Dunlop in 1911. Under the guidance of Richmond architect Duncan Lee, the couple made far-reaching changes to the house. Most conspicuous was the new, steeply sloping roof of the main house, seven feet higher than its predecessor. Under this new roof were three additional bedrooms, two baths, and a sizable sitting room, all illuminated by new dormer windows.

To maintain the scale of the two flanking dependencies in relation to the enlarged mansion, Lee rebuilt their river-front walls, adding nearly ten feet to the depth of each outbuilding. In each case, the result was a higher roofline more in keeping with the increased height and scale of the main house.

The single Bisland connector was demolished and the flanking dependencies now connected to the main house by gallery-like "hyphens." These new connectors created additional living and service spaces while providing access to the dependencies both from the front and the back rooms of the main house.

Above left. Mollie McCrea, photographed about the time of her marriage to Archibald McCrea in 1911. *Courtesy, Charles McCrea.*

Above right. Archibald McCrea as a young man. *Courtesy, Charles McCrea.*

Detail. The old laundry as it appeared before the work began.

Right. The land front following Mr. and Mrs. McCreas' restoration. *Courtesy, Ann Gray.*

Members of the McCrea family enjoy lunch during a visit to inspect the progress of the work. The Booth-era porch is not yet dismantled. *Courtesy, Joseph Tilden.*

Left. Second-floor rooms in the midst of renovation.

Right. The Bislands' connecting gallery and the adjoining cellar entry were entirely rebuilt.

The continuing importance of the past—photographers consult with Mrs. McCrea as they prepare to copy a portrait.

The McCreas' restoration was part of a larger happening that had overtaken Virginia in the decades after 1900. During this period, the state experienced a rebirth of sorts—a new infusion of energy and wealth—as scores of affluent Northerners, some claiming ancestral ties to the Old Dominion, acquired, reconditioned, and modernized Virginia's ancient plantation homes. Between 1925 and 1935 alone, more than a dozen major houses were restored as growing numbers of properties appeared on the real estate market. Berkeley, Brandon, Carter's Grove, Claremont, and Tuckahoe, all situated on the James River, were among the most conspicuous of these restorations undertaken by private owners.

The reclamation of Virginia's old mansions owed much to a growing fascination with the genteel style of country living. The English squire was the obvious model for America's new rural elite, and the old plantation house became the inevitable habitat for this way of life.

Carter's Grove afforded rural pleasures for family and guests alike. *Courtesy, Joseph Tilden.*

Archibald and Mollie McCrea in 1936. *Courtesy, Dementi Studios.*

Balanced against the pleasures of rural living was the conviviality for which the McCreas and their peers were noted—the essential ingredient of a grand lifestyle for which Carter's Grove and other Virginia mansions were remade. Such restorations embodied the emergence of a new rural gentry with its love of entertainments, its growing taste for the pleasures of

country life, and its cultural identification with England.

Mr. and Mrs. McCrea thus transformed one of Virginia's great colonial mansions into one of America's great country houses. Here they perfected what a contemporary writer called "the art of living in the country." Photographs from the 1930s reveal Mr. McCrea's abiding interest in horses and riding, while Mrs. McCrea is said to have been an accomplished gun who frequently enjoyed shooting with friends and guests.

Archibald and Mollie McCrea imprinted Carter's Grove with their own special qualities, and the story of the mansion in the twentieth century must be, in large measure, their story. Nevertheless, the real significance of Carter's Grove eludes us if we fail to see its restoration in the context of contemporary trends and currents of thought represented by the restoration of Virginia's grand homes, the gentrification of country life, and the growing interest in America's past as a bulwark against unsettling change.

Above. Riding was a favorite pastime. *Courtesy, Charles McCrea.*

Right. Mr. McCrea and daughter Ada treat visitors to refreshments under the shade of a tree. *Courtesy, Charles McCrea.*

Left. Ada McCrea listens as conversation mingles with the sounds of a local guitarist performing on the periphery of the group. *Courtesy, Charles McCrea.*

Below left. Mrs. McCrea and Nelson Rockefeller pause for a photograph on the river-front steps of Carter's Grove. *Courtesy, Charles McCrea.*

Below right. Mrs. McCrea takes a moment to enjoy the beauty of a densely planted flower bed. *Courtesy, Charles McCrea.*

After the death of Archibald McCrea in 1937, Mollie McCrea continued to entertain at Carter's Grove often until her death in 1960. In her will, she stated a long-standing "hope and ambition" that the property might be maintained for the benefit of subsequent generations.

Her wish coincided with the vision of Winthrop Rockefeller, the son of John D. Rockefeller, Jr., Colonial Williamsburg's benefactor. Through Winthrop Rockefeller's leadership as chairman of Colonial Williamsburg's Board of Trustees, the Sealantic Fund, a Rockefeller family philanthropic organization, purchased the property from Mrs. McCrea's estate. In 1969 Carter's Grove was transferred by deed of gift to the Colonial Williamsburg Foundation.

Today, with further generous support provided by the trustees of the Winthrop Rockefeller Charitable Trust, Carter's Grove is open to the public, presenting a sweeping panorama of Virginia's past that embraces more than four hundred years of human endeavor.

Carter's Grove, 1956.

The Mansion

*C*rowned with a steeply pitched roof and connected with its dependencies into a single elongated structure, Carter's Grove bears a striking resemblance to Westover, the James River mansion of the Byrd family. Westover's fame centered on the memorable character of William Byrd II, a polished and highly educated man of the world. Planter, lawyer, naturalist, and bibliophile, Byrd moved in prominent circles on both sides of the Atlantic and left a series of diaries that have become the foundation of his literary reputation.

Westover and the Byrds epitomized gentry life in colonial Virginia, and for that reason held a special attraction for Mr. and Mrs. McCrea, who wanted to buy the estate. When their wishes could not be realized, they set out to remake Carter's Grove in the image of Westover. Once the work was complete, they decorated their new house with important objects acquired at the 1919 sale of Westover furnishings. As restored, Carter's Grove is an evocative manifestation of Colonial Revival style applied to the arts of architecture, interior decoration, and landscape design.

The river front of Westover as restored by Mrs. C. S. Ramsay in 1898.

The growing importance of the automobile prompted Mr. and Mrs. McCrea to create an oval drive on the land-side approach. It was based on an example at Castle Hill in Albemarle County.

Hall

$\mathscr{B}$ecause eighteenth-century Virginians often regarded the river side of the house as the front, the most elaborately finished rooms were usually located there. This is the case at Carter's Grove, where approaching visitors see the best rooms and the greatest number of windows on the river front.

The stair is the focal point of the hall, framed like a picture by the magnificent elliptical arch. Despite the room's formality, it probably was used most intensively during the summer months when the routine of decorous manners and dress was suspended. Here sat the planter with his doors open and his wig off, hoping to benefit from an occasional gust of air. The servants who were not immediately occupied in some household task may have waited in the adjoining passage.

A dozen or so chairs, tables for tea or dining, and a couch for noontime naps would have served the needs of family and visitors alike. Like the "saloon" at Colonel Thomas Mann Randolph's house, Tuckahoe, the hall may have functioned as "an occasional ballroom" as well, where the Burwells and their guests would have danced during public entertainments.

In the McCrea era, the hall provided an impressive introduction to the house and the people who lived here. Above the tables that flank the arch are a pair of seventeenth-century paintings, called "Peace" and "Plenty" by the McCreas, which are believed to have been painted by the Neapolitan artist Luca Giordano (1632–1705) or one of his pupils. Portraits of John Minge and of his wife, Sarah Harrison Minge, originally of Berkeley in Charles City County, hang in the passage. These portraits came to Mrs. McCrea through her first marriage to Petersburg tobacconist David Dunlop.

The magnificent finishes of the stair were created by a highly skilled team of eighteenth-century tradesmen. The floral ornaments below each tread were executed by the same carver who had embellished stairs at Rosewell, the Gloucester County seat of the Page family, and Tuckahoe, a Randolph house in Goochland County. The balusters and newel post were turned by a man identified only as "Sumpter" in Burwell family records, possibly a slave. By the time of the McCreas' arrival, the clock on the stair had become a standard device in Colonial Revival interiors.

Adorning the passage are portraits of John and Sarah Minge of Weyanoke, ancestors of Mrs. McCrea's first husband, David Dunlop. Mrs. Minge's brother was William Henry Harrison, ninth President of the United States. The resplendent canvases in the hall were known to the McCreas as "Peace" and "Plenty." They have since been identified as "An Allegory on Monarchy" (illustrated) and "An Allegory on Sacred and Profane Love." Architect Duncan Lee cleverly used tasseled bellpulls to activate the electric lights. A tray of calling cards evokes the numerous guests entertained at Carter's Grove, and the vanishing bit of etiquette they observed in leaving a memento of their visit.

Dining Room

*T*his room served as the Burwells' principal chamber or bedroom where the couple shared intimate moments with their children and with one another. The most private of the first-floor rooms, it was appointed rather simply compared to the river-front rooms.

Because the McCreas planned to entertain frequently, they devoted the entire ground floor of the original house to public uses. Thus, what had been in the Burwells' time an inner sanctum of family life now became a vortex of hospitality. To facilitate this change, the room

was enlarged by removing a closet and vestibule at its eastern end. Here Mr. and Mrs. McCrea dined with their guests on formal occasions. Family portraits include likenesses of Mr. McCrea's mother and father and a painting of Mrs. McCrea copied from a miniature made during her honeymoon.

From her vantage point above the fireplace, Archibald McCrea's mother presides over a formal table laid with a cutwork cloth and silver candelabra. The Sheffield plate epergne and the flamboyant wine trolley, both inherited by the McCreas' daughter, Ada, have recently returned to Carter's Grove. The epergne is an English piece dating from about 1815. The trolley is somewhat later, and is probably Continental in origin. Over the serving cabinet is a portrait of Mollie McCrea as she appeared at the time of her first marriage to Petersburg tobacconist David Dunlop in 1896.

Drawing Room

Although Mr. and Mrs. McCrea referred to this space as their drawing room, it really functioned most often at mealtimes. Mrs. McCrea occasionally hosted luncheons attended by a few friends in this room, but it was usually the McCreas themselves who gathered here for meals. On public occasions when the number of guests present for dinner made it impractical to serve everyone in the dining room, the drawing room, which was convenient to the kitchen and pantry, could function as an overflow dining area.

In the eighteenth century, the Burwells had also used this space as a dining room. Its social importance in that capacity is reflected in the robust architectural appointments. On the chimney breast just below the cornice are emblems of the earth's bounty—wheat, garlands, fruit, and flowers—all symbolic of the Burwells' unstinting hospitality.

Breakfast Room

*M*r. and Mrs. McCrea usually break-fasted in this room, enjoying the view of the James River as they ate. A series of blind arches painted to give the masonry an antiquated appearance adds to the visual interest of the room. At the far end, a large mirror doubles the apparent size of the space. The masonry walls and clay tile floor present a striking contrast to the rich furnishings and tie the room to the outdoors.

Butler's Pantry

On the occasion of formal dinners, this room served as a staging area where food was brought from the kitchen, arranged in serving dishes, and taken to the dining room. After dinner the dishes were washed and returned to the cupboards.

This doorstop is made of cast iron.

Mrs. McCrea arranged flowers from her garden in the pantry.

Kitchen

Originally detached from the main house, this space has functioned as a kitchen since 1740. When the McCreas resided at Carter's Grove, even the kitchen became a showplace of sorts. To reduce the impact of modern amenities on the room's historical ambience, a scullery, pantry, and servants' hall were placed in the added space beyond the south wall. Originally they were accessible through a pair of broad doorways. In the kitchen proper, a sink was hidden below the countertop on the north wall.

The immense fireplace was a fictional creation intended to evoke the huge fireplaces of New England with their connotations of hospitality and domestic comfort. Dozens of cooking implements intended more for show than for use impart a theatrical quality to the setting where Mr. and Mrs. McCrea hosted informal gatherings over roast oysters or Brunswick stew.

A Hotpoint electric oven and range suggests the variety and extent of kitchen tasks necessary to sustain the McCreas and their guests. In a corner of the adjoining servants' hall, lunch awaits housekeeper Edna Washington, a reminder of the practice, then prevalent, of separating domestic servants from the family. Antique cooking utensils—copper pots and pans, crockery, and old bottles—contribute to the "historical" atmosphere of this workaday space.

Maid's Room

$\mathcal{S}$ ituated over the kitchen, this room was first earmarked for guests, but later it was reserved for the use of live-in housekeeper Edna Washington. No references to prior ages here, no attempts to ally oneself with history. In this room, Washington spent the hours she called her own, close by yet separated from the family she served. One wonders to what extent she drew her own identity from the employers she served—in what measure she entertained proprietary feelings for the place to which she devoted a significant portion of her life.

Red Bath

$\mathcal{W}$herever possible, the McCreas and architect Duncan Lee sought to move modern services and amenities such as the heating plant, bathrooms, kitchen, servants' rooms, and family spaces to the extremities of the complex—to the hyphens and dependencies. This preserved the historic ambience of the main house and allowed the architect to distribute plumbing, heating, and electricity with greater efficiency.

Bound elsewhere by the proprieties of period decor, architect Lee finally "cut loose" in

this modern, backstage space, using boldly figured black-and-white marble in concert with chrome fittings and red textiles. The toilet, however, was another matter—it was hidden behind a pair of raised-panel doors, masquerading as a linen closet. In a time when many Americans still relied on outdoor plumbing, the amplitude and comfort of this bathroom would scarcely have been imaginable.

Guest Room

$\mathcal{T}$his chamber, which looks out over the James River, was one of Nathaniel Burwell's best rooms and may have been reserved for company. The McCreas certainly used it in this way so that visitors enjoyed direct access to the luxurious bathroom next door. Together, this pair of spaces formed an opulent guest suite.

The emergence of a national highway system in the 1920s and '30s greatly improved the accessibility of Carter's Grove to friends and strangers alike. Interior Secretary Harold Ickes, a frequent visitor, slept here on numerous occasions. Other guests included Nelson Rockefeller, Gertrude Stein, and Walt Disney. Riding apparel on the wardrobe suggests the amusements available to those who enjoyed the McCreas' hospitality.

Office

$\mathcal{E}$arly in the century, the Bisland family furnished this room as an office and library. The McCreas continued this arrangement, creating the built-in bookcases that now flank the arch. The desk is fitted with a typewriter and a telephone, modern conveniences that allowed men in McCrea's era to live as country gentlemen while maintaining contact with their businesses in the city.

The collection of books reflects Mr. and Mrs. McCreas' interest in history (Virginia history in particular), travel, gardening, hunting, and fiction. Several volumes bear the bookplates of Archibald McCrea's father or grandfather.

Just how the Burwells used this space remains unclear. The great arch suggests a public function of some importance—perhaps an upstairs drawing room, or a place in which to have supper, the light meal served at the end of an evening's entertainment.

Works of fiction and history in the bookcase profile the family's literary interests. Over the antique clothespress is a nineteenth-century print of "Lady Washington's Reception," an example of the historical consciousness that directed the McCreas' decoration of their home. On the table between the tufted armchairs, a portrait of Archibald and Ada McCrea captures the affection that characterized their relationship. By the door is a photograph of tennis star, author, and army officer Helen Hull Jacobs, a frequent and favored guest.

Mr. McCrea's Bedroom

*T*his room served the Burwell family as a bedchamber. Situated on the river front of the house, it was probably regarded as one of the better upstairs rooms, though it is no more elaborate than the other second-floor spaces. The architectural consistency of the upper floor contrasts markedly with the varied appointments of the lower rooms, which were intended to denote the distinct function of each space.

This was Archibald McCrea's bedroom. Smaller and less flamboyant than Mrs. McCrea's room, this airy, sunlit space offered a respite from the formality of the public rooms below.

Brass harness ornaments above the fireplace are reminders of McCrea's fondness for equestrian pursuits. At his suggestion architect Lee included a stable and paddock in his plans for the restoration of Carter's Grove.

Mrs. McCrea's Bedroom

*M*rs. McCrea enjoyed the use of this bedroom, the largest at Carter's Grove. The bed in the Chinese taste, a nineteenth-century reproduction, is one of many pieces in the house purchased from the Westover furnishings of Mrs. C. S. Ramsay.

In the eighteenth century, the far end of this room was cordoned off to form a pair of closets. In 1907, the Bisland family converted these closets to a bathroom by removing the partition between them. Later on, the McCreas pulled out the remaining bathroom wall, moving toilet facilities into the new hyphens.

This portrait is of Mrs. McCrea and her mother, Charlotte LaMoine Johnston.

Renowned in her youth as a woman of striking beauty, Mollie McCrea completed her toilet in front of the large nineteenth-century mirror installed above her dressing table. The use of chintz at the windows, around the bed, and on the chaise longue was typical of the decorative fashions of the day. As revivals of earlier patterns, these fabrics were essential to the idea of a colonial interior. Chintzes were popular in English country houses of the period, and were thus associated with rural gentility as well.

Stair Passage

A high degree of workmanship is evident in the original stair and also in the later flights that now continue to the third floor. The added steps provide access to the new upstairs rooms created for the McCreas. A carefully crafted joint in the handrail reveals where the old and new work meet. The wisdom of such changes is debatable, but one can only admire the skill with which they were carried out. Above the landing a protruding ledge marks the transition between the masonry walls of the first and second stories.

On the railing of the lower flight is a series of gashes, reputedly the work of Banastre Tarleton, commander of the British cavalry in this area in 1781. Traditional accounts portray Tarleton hacking on the banister with a saber while riding his horse up the stair. This legend seems to have gained currency by the end of the nineteenth century, and has been dismissed by later historians. However, the recovery of a brass ornament bearing Tarleton's family crest by archaeologists working at Carter's Grove in the 1970s suggests that Tarleton was at least here even if he did not attack the staircase.

Just above the landing, a difference in wall thicknesses of the first and second floors creates a ledge on which to display a collection of brass candlesticks, a reference to the supposed practice of "lighting" people to their rooms. From the second floor, a new flight of stairs leads to the rooms Mr. and Mrs. McCrea created on the level above. Here, under the mansion's enlarged roof, was daughter Ada's room, plus several guest bedrooms and a storeroom.

Library

This land-front room probably served the Burwell family as a private parlor, but the McCreas designated it for public use as a library. It had long been the custom for men and women to separate after meals, the ladies moving to a drawing room, the men to a library or smoking room where they indulged in alcohol or some form of tobacco before rejoining the ladies. Perhaps Mr. McCrea invited male companions to join him in the library after dinner for conversation, a cigarette, and a drink. The dark richness of the furnishings, textiles, and bookbindings may have been intended to imbue the room with a masculine quality.

Interior Secretary and family friend Harold L. Ickes introduced the McCreas to President Franklin D. Roosevelt and involved them in the work of his admin-

istration. Through Ickes, Mr. McCrea was appointed to the Interior's "Advisory Board on National Parks, Historic Sites, Buildings and Monuments." As a board member, McCrea immersed himself in the nation's growing historic preservation movement, advising on the government's role in the care and interpretation of America's past. He was well positioned to serve, having restored a great eighteenth-century house and living but a short distance from Williamsburg and the National Park Service installations at Jamestown and Yorktown.

Above. Secretary of the Interior Harold L. Ickes.

Left. The likeness over the fireplace depicts a youthful, ebullient McCrea ancestor. The original, supposedly painted by Thomas Sully, was here copied by Frank B. A. Linton. Linton produced several other portraits for the mansion. Family photos on the marble-top bookcase show Archibald and Mollie McCrea and their daughter, Ada. Mrs. McCrea's funeral took place in this room.

The massive Gothic bookcases with their ivory accents were purchased at the Westover sale of Mrs. Ramsay's furnishings. Their contents add to the rich appearance of the room. The desk, covered with leather and veneered in burled walnut, is probably Continental and is noteworthy for having drawers on both fronts. The photograph is inscribed to the McCreas "from their friend Franklin D. Roosevelt."

Refusal Room

The Refusal Room takes its name from the alleged rejection of two famous suitors, George Washington and Thomas Jefferson, in this space. These traditions had long been associated with Carter's Grove, but it seems to have been Mrs. McCrea who attached them to this particular room. The romantic chronicles recounted here were as necessary as bricks and mortar to the McCreas' remaking of Carter's Grove.

To underscore the theme of legendary romance, Evelyn Byrd's thwarted love affair with the Earl of Peterborough was added to the stew. In spite of the Earl's earnest suit, the story goes, the girl's father, William Byrd of Westover, would not consent to the union. Denied the hand of her suitor, Evelyn died of a broken heart. From his vantage point above the fireplace, the Earl (garnered at the Westover sale) gazes across the Refusal Room at the heartbroken Evelyn. Her ghost now inhabits the room, it is said, and a bouquet of carnations left in the room overnight will be pulled apart and scattered about by morning. Unable to purchase Westover, the McCreas settled for one of its ghosts!

Nathaniel Burwell probably used this space as a parlor or drawing room, furnishing it with his best furniture and textiles. Here visitors exchanged pleasantries with the family, read aloud, played cards, or enjoyed music. Ladies probably took their after-dinner tea here as well, leaving the gentlemen to their bottle in the dining room.

The architectural appointments of the room emphasize its social importance in the eighteenth century. The lower portion of the white marble chimneypiece has projecting shoulders, or "crossets," and is the only one of its kind in the house. The upper portion, with its frieze of Siena marble, may have been added late in the eighteenth century. As in the dining room, a classical enframement adorns the chimney breast.

Richly figured silk textiles and an Aubusson rug identify this as the McCreas' best room—a status it has enjoyed since the time of Carter Burwell. The upper section of the chimneypiece was probably added by Burwell's son and heir, Nathaniel. The Earl of Peterborough hangs above the mantel.

In this space the McCreas gathered romantic icons of Virginia's past—George Washington, the Earl of Peterborough, and, between the windows, his sweetheart, Evelyn Byrd. Like the legend of Evelyn's thwarted love affair with the Earl, many objects in this room, including the carved side chairs by the windows, came to Carter's Grove from Westover.

New Sitting Room

*T*his twentieth-century room joins the main house to what was once a freestanding laundry. The McCreas frequently gathered here with guests for a game of cards or to enjoy drinks before dinner. Light and airy in its decor, it was one of the mansion's most comfortable living areas. One can imagine the arched doors of both fronts standing open on a summer day to admit breezes from the river. In cooler weather, sunlight streaming in through the transparent south wall would have bathed the room in a warm glow.

The brighter palette of this room imparts an up-to-date feeling to the space, as do the taut lines of the sofa and looking glass. Flanking the glazed doors are a series of tinted hunting prints that invoke the life of the English squire and its latter-day re-creation on the James River. Ivy plants to either side of the sofa echo the floral patterns of the draperies, suggesting a link to the outdoors. Six sets of doors, all exiting at ground level, make this connection real.

The portrait of President William Henry Harrison is an early copy from the original by Rembrandt Peale. The children from Mrs. McCrea's first marriage were related to Harrison through their father, David Dunlop. By an open door, the card table stands ready for a game of backgammon. The shield-back chairs were probably made in Baltimore at the end of the eighteenth century.

This three-part chest, which dates from about 1725, was adapted in order to display dessert china on the two upper shelves and a tea set on those below. On top is a garniture of Chinese ginger jars. The compartment below the china display incorporates a fall-front desk.

Smoking Room

Known in the 1930s as Mr. McCrea's Room, this was where the McCreas spent most of their time. Architect Duncan Lee created the space by enlarging the separate laundry and removing the partition that once bisected its lower floor. The resulting room was the largest and least formal of the major ground-floor spaces, a counterpoint to the formal propriety of the main house. Braided rugs, Windsor chairs, oak furniture, antique copper vessels, and wrought-iron fireplace equipment all contribute to its comfortable domestic character. Around the walls are reminders of Mrs. McCrea's blood ties with Virginia's illustrious past—Spotswood portraits, a coat of arms, and an engraving of Revolutionary War hero Peter Francisco.

The half-circular table was a favorite of Mrs. McCrea's where she played solitaire or surrounded herself with friends and admirers. Nothing in the house illustrates more aptly the role Mollie McCrea cultivated for herself at Carter's Grove.

The half-circular piece of furniture was originally called a "drinking table." The open end was placed against the fireplace, and gentlemen might warm their feet at the fire while warming their insides with spirits.

The photograph beside the telephone commemorates a 1924 trip to see the Sphinx and the great pyramids at Giza.

Gardens and Grounds

Toward the river, one can see evidence of Mr. McCrea's abiding interest in the landscape. Working closely with landscape architect Arthur Shurcliff, McCrea traveled extensively in Virginia, visiting old estates in search of ideas and plant materials to be employed at Carter's Grove. While the results of their work are most evident on the land side of the house, Shurcliff and McCrea did reopen the vista toward the James River. Looking out from the privacy of his second-floor bedroom, McCrea must have taken great satisfaction in this decision.

Lying at the foot of terraces laboriously cut out in the 1740s, the eighteenth-century garden has been re-created on the original site identified by archaeological excavations in the 1970s. The present enclosure and paths replicate an arrangement dating from the time of Nathaniel Burwell. Garden descriptions and plant lists of the period suggested a mixture of ornamental and utilitarian plants that seems unusual to modern eyes. Here the Burwells walked to pass the time, to savor the air, or to enjoy one another's company.

On the land-side approach, Shurcliff allowed Edwin Booth's grove of locust trees to remain, but set out an "irregular thicket of evergreen planting . . . arranged in scattered formation to look like overgrown planting of great age." Just as the collected lore of the Refusal Room threw the mansion's past into higher relief, this mature shrubbery gave Carter's Grove a suitably venerable appearance.

Landscape architect Arthur Shurcliff created an ornamental setting with plant materials appropriate to a country estate. The stable and paddock reflect the family's interest in horses and riding.

Slave Quarter

*T*he quarter was reconstructed in 1989 on the site of the houses occupied by slaves who lived at Carter's Grove in the decades immediately before and after the Revolutionary War. These were people who labored in the fields nearby and worked at skilled trades like carpentry and coopering. Below the largest house archaeologists found about a dozen cellar pits filled with objects like ceramics and tools reflecting the lives of the occupants and suggesting that it may have once housed upward of twenty people. Additional, smaller houses illustrate more segregated accommodations and the formation of African-American families at the site.

Presentation of the quarter began in 1990 by populating it with a hypothetical group of people whose names and relative ages were drawn from the Burwell records. Research on the lives and lineage of African-American slaves is developing rapidly, and recent work is now defining a much more precise picture of the identities and origins of Burwell slaves. Some were descended from African-Americans brought to York County to work alongside white indentured servants as early as 1666, while others had African parents and grandparents from the Niger River Delta and were bought by Robert "King" Carter directly off slave ships in the Rappahannock River.

The design of the quarter was developed from nearly a decade of documentary research and study of surviving buildings in the Chesapeake as well as from archaeology. The result is a graphic portrayal of the material circumstances of ordinary people in eighteenth-century America. But this is more than a portrait of simple life in the preindustrial past. The small collection of cheaply built Anglo-American houses is grouped with yards, garden enclosures, and ways of using objects that are distinctly African-American. The quarter confronts the economic decisions made by Burwell and his counterparts as well as some of the means by which black people resisted the system of slavery. Visitors to the slave quarter at Carter's Grove gain some understanding of how race relations evolved in eighteenth-century America and are encouraged to consider how that past affects us today.

Nighttime obscured many of the quarter's ties to the larger economic landscape and set the scene for vigorous expressions of community.

Little was known about the lives of slaves in eighteenth-century Virginia until the last two decades when historians began to probe the distinctive nature of their culture and archaeologists began to excavate sites such as this. From the arrangement of daily activities across the site to the pits in which people stored their possessions, the re-created Carter's Grove quarter provides a physical portrayal of African-Americans who lived on Nathaniel Burwell's home farm.

Wolstenholme Towne and the Winthrop Rockefeller Archaeology Museum

Exhibits at the museum include many artifacts recovered from the site of Wolstenholme Towne, together with comparable seventeenth-century objects from across Europe.

*I*n the Winthrop Rockefeller Archaeology Museum, exhibits and artifacts recovered from nearby sites tell the story of how Colonial Williamsburg archaeologists discovered Martin's Hundred and its administrative center, Wolstenholme Towne, in the 1970s. Photographs show archaeologists exploring and excavating the site. Parts of weapons, agricultural tools, ceramics, and domestic artifacts illustrate how the settlers lived in the Old World and their strategies for survival in the New World. Paintings, documents, and audiovisual exhibits explaining how archaeologists assess their discoveries add an important element to the story. Two helmets, the first intact close helmets found in North America, are among the important artifacts on display in the Winthrop Rockefeller Archaeology Museum.

Adjoining the museum is the site of Wolstenholme Towne. A schematic interpretation of part of the settlement includes the Wolstenholme Towne fort, store, barn, and domestic unit. Audiotapes and partial reconstructions of these buildings evoke the seventeenth-century life and landscape of that community.

Revered for its haunting and serene beauty, Carter's Grove illuminates the history of the New World like no other place. From the earliest occupation by native peoples to Mr. and Mrs. Archibald McCreas' embellishment of a great American house, every period, every site has something important to tell us about broader aspects of the American story. At Wolstenholme Towne, English and native cultures collided as European aspirations were acted out on the American continent. The slave quarter evokes the distinctive character of African slavery in this new land, while Carter Burwell's mansion embodies the emergence of a gentry class in Virginia and the rise of a transatlantic community of consumers. Edwin Booth's rescue of Carter's Grove in the years following the Civil War reflected a new historical awareness in America, born of the reconciliation between warring states and culminating in the McCreas' confident elaboration of the property in this, the "American Century." At Carter's Grove, these themes assume shape and substance, weaving themselves into a vivid chronicle of American life.

Acknowledgments

This book represents the work of many people at Colonial Williamsburg and elsewhere. I am indebted to Lawrence Henry for reading my manuscript and for valuable suggestions on the text and illustrations. Larry has done more than anyone to put the McCreas back in the house. No one knows more about this family; no one has approached the topic more thoughtfully.

Ruth Rabalais also read the manuscript and saved me from several embarrassing errors. Louise Kelley provided essential information on numerous McCrea objects. I have relied heavily on the research notes of Diane Dunkley, who was the first to bring modern curatorial methods to the assessment and care of the McCreas' belongings. David Muraca summarized what is known about Native Americans at Carter's Grove, and his ideas became the foundation for that section of the text.

I would also like to thank photographers Kevin Burke, Tom Green, and Dave Doody, flower arranger Peg Smith, and designer/stylist Sue Rountree for bringing Carter's Grove to life visually and Mary Keeling and the staffs in the Foundation's audiovisual library and audiovisual laboratory for their assistance with photographs. Thanks too to book designer Helen Mageras and senior editor Donna Sheppard in the Department of Publications for their help.

Finally, I want to acknowledge the contributions of Ed Chappell, director of architectural research. As the prime mover behind the Carter's Grove slave quarter, he agreed to write that portion of the text. His insights reflect years of study in the architectural and documentary record. Ed was instrumental in bringing me to the project.